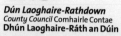

Our World

Gas

By Sarah Levete

Aladdin/Watts
London • Sydney

© Aladdin Books Ltd 2006

Designed and produced by
Aladdin Books Ltd
2/3 Fitzroy Mews
London W1T 6DF

First published in 2006 by
Franklin Watts
338 Euston Road
London NW1 3BH

Franklin Watts Australia
Hachette Children's Books
Level 17/207 Kent Street
Sydney NSW 2000

A catalogue record for this
book is available from the
British Library.

ISBN 0 7496 6279 4

Printed in Malaysia

All rights reserved

Editor:
Katie Harker

Design:
Simon Morse
Flick, Book Design and Graphics

Consultants:
Jackie Holderness – former Senior Lecturer
in Primary Education, Westminster Institute,
Oxford Brookes University

Rob Bowden – education consultant, author
and photographer specialising in social and
environmental issues

Illustrations:
Ian Thompson

Picture researcher:
Alexa Brown

Photocredits:
l-left, r-right, b-bottom, t-top, c-centre, m-middle. Front cover, 01, 3tl, 3mtl, 3mbl, 5tl,
5br, 6tl, 6bl, 10bl, 11tl, 12tr, 13tr, 13bl, 15tl,
15br, 16tl, 16ml, 18tr, 20tl, 21br, 23tr, 23bl,
24tl, 26tr, 27tr, 29tl, 29br, 30tr, 31bl –
www.istockphoto.com. Back cover, 2-3, 5bl,
22tr, 22bl, 30br – Comstock. 3bl, 18bl, 22ml –
Corel. 4ml, 14tl – Select Pictures. 4bl, 14bl –
© Calor Gas Limited. 5ml, 11br, 12bl, 30mr –
Hydrowingas. 5tr, 10tl, 19br, 20bl, 28bl, 31tl
– US Department of Energy. 7tr – TongRo
Image Stock. 8tr, 24bl – Stockbyte. 8bl – New
Mexico Tech. 9tr – Captain Albert E.
Theberge, NOAA Corps. 9bl – Corbis. 16bl,
16bm – Ingram Publishing. 17tr, 25tl, 28tl –
Digital Vision. 17m, 17br – Sanyo Air
Conditioning. 21tr – FEMA. 25br – Robert S.
Waselwski, Tampa Electric Company. 26bl,
27bl – Photodisc.

CONTENTS

Notes to parents and teachers

This series has been developed for group use in the classroom as well as for children reading on their own. In particular, its differentiated text allows children of mixed abilities to enjoy reading about the same topic. The larger size text (A, below) offers apprentice readers a simplified text. This simplified text is used in the introduction to each chapter and in the picture captions. This font is part of the © Sassoon family of fonts recommended by the National Literacy Early Years Strategy document for maximum legibility. The smaller size text (B, below) offers a more challenging read for older or more able readers.

Using natural gas

Natural gas gives off lots of energy when it burns. Energy makes things work. We use gas to heat water for washing.

A

◀ At home your gas comes from a pipe in the street or from a tank.

In towns and cities, most houses are supplied with gas from 'the mains'.

B

Questions, key words and glossary

Each spread ends with a question which parents and teachers can use to discuss and develop further ideas and concepts. Further questions are provided in a quiz on page 30. A reduced version of pages 30 and 31 is shown below. The illustrated 'Key words' section is provided as a revision tool, particularly for apprentice readers, in order to help with spelling, writing and guided reading as part of the literacy hour. The glossary is for more able or older readers. In addition to the glossary's role as a reference aid, it is also designed to reinforce new vocabulary and provide a tool for further discussion and revision. When glossary terms first appear in the text, they are highlighted in bold.

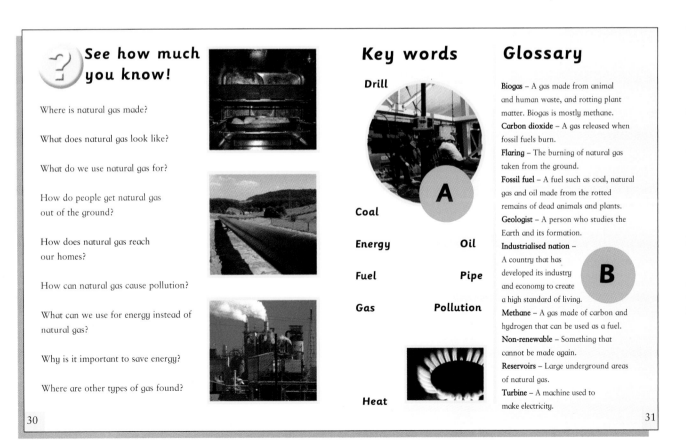

See how much you know!

Where is natural gas made?

What does natural gas look like?

What do we use natural gas for?

How do people get natural gas out of the ground?

How does natural gas reach our homes?

How can natural gas cause pollution?

What can we use for energy instead of natural gas?

Why is it important to save energy?

Where are other types of gas found?

Key words

Drill

Coal

Energy   Oil

Fuel   Pipe

Gas   Pollution

Heat

Glossary

Biogas – A gas made from animal and human waste, and rotting plant matter. Biogas is mostly methane.

Carbon dioxide – A gas released when fossil fuels burn.

Flaring – The burning of natural gas taken from the ground.

Fossil fuel – A fuel such as coal, natural gas and oil made from the rotted remains of dead animals and plants.

Geologist – A person who studies the Earth and its formation.

Industrialised nation – A country that has developed its industry and economy to create a high standard of living.

Methane – A gas made of carbon and hydrogen that can be used as a fuel.

Non-renewable – Something that cannot be made again.

Reservoirs – Large underground areas of natural gas.

Turbine – A machine used to make electricity.

30

31

What is natural gas?

Natural gas is invisible. It has no colour, shape or smell. But when gas burns, a flame gives off heat and light. Natural gas is a fossil fuel made from the remains of plants and animals that lived millions of years ago.

◀ **Natural gas is an important source of light and heat energy.**

Natural gas is made up of different chemicals and gases. The main ingredient is **methane**, a gas that burns easily. When we burn natural gas, it gives off a lot of energy. This energy is used for cooking and heating. We can even use it to power cars.

▶ There are many gases in the air.

Most of the air that surrounds us is made up of the gases oxygen and nitrogen. We need the gases in the air to breathe. Natural gas is a mixture of different gases and chemicals that form underground.

Natural gas formed millions of years ago from the rotten bodies of dead animals and plants.

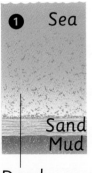

❶ Sea

Sand
Mud

Dead animals and plants sink down.

Over time they turn to oil and gas.

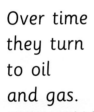

❷ Oil and gas
Rock

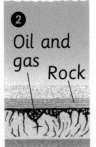

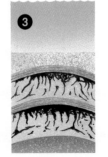

❸

Oil and gas collect in layers under the sea.

Over millions of years, dead plants and animals sank deep into the earth or the seabed. They were pushed down by soil and water. Some remains were squeezed at high temperatures to form rocks. Lower down, oil and natural gas formed.

 How many fossil fuels can you name?

Where do we find natural gas?

Natural gas forms deep under the ground or under the seabed. It is usually found near oil. Large areas of natural gas are found all over the world, from Nigeria to Russia.

◀ **Scientists search for gas in rocks and soil.**

Small amounts of natural gas rise up from the ground and escape through tiny holes in rocks and soil. The gas disappears safely into the air. But, when the gas meets rocks without any holes, it becomes trapped. **Geologists** use a geophone (left) to scan and record movements underground.

▶ Special machines show where gas can be found.

When natural gas becomes trapped it forms large areas called **reservoirs**. Geophone recordings can be shown on a computer screen to work out what types of rock are underground and whether there are areas of trapped gas.

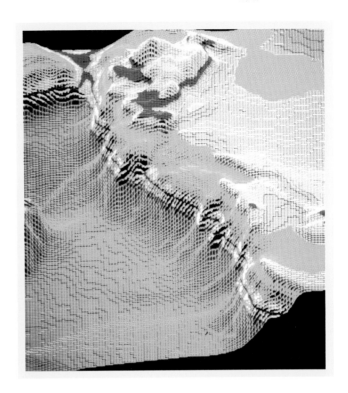

Lightning can cause natural gas to catch fire.

Thousands of years ago, people saw fires coming from the ground. They believed that these fires were mysterious signs from the gods. However, they were probably started when lightning set alight natural gas that had seeped out of the earth.

What are fossil fuels used for?

Drilling for gas

Have you ever dug a hole in some sand or in the earth? How far could you dig? To reach large areas of natural gas, people drill deep under the earth. The drill makes a hole (or a well) in the solid rock that surrounds the natural gas.

◀ This platform is drilling for gas in the middle of the sea.

There is a lot of gas deep beneath the seabed. A gas rig is built and then transported to the middle of the ocean. This is called an offshore rig. From here, drillers are able to reach the depths of the sea. They live for months at a time on the rig. Some rigs are the size of villages!

◀ **Natural gas is cleaned before it reaches our homes.**

When a drill hits an area of natural gas, the gas collects in a well and is sent along a pipe to a processing plant. Here, any unwanted gases or chemicals are removed. These can be used later for other products or fuels.

Drilling for natural gas can harm the natural world.

Huge drilling rigs and massive pipes spoil the beauty of the natural world. Pipelines can destroy the habitats of plants and animals, and the chemicals and fuels can upset the natural balance of the area.

 What might it be like to work on a gas rig?

Transporting natural gas

Gas can be taken from the ground and sent through a network of pipes to homes around the world. Gas can also be carried in metal containers for camping equipment or portable heaters.

◀ **Some gas pipes are as wide as a person.**

A country with large natural gas deposits, such as Russia, sends the fuel under the sea and overland to other countries. In Norway, the 1,200 kilometre Langeledd pipeline transports gas to the UK along the seabed. Gas usually travels at about 24 kilometres an hour along these large pipelines!

▶ Liquid gas can be moved in a lorry.

Chilling gas turns it into a liquid. This takes up less room than its gassy original form. It is easier and cheaper to transport liquid natural gas to areas that cannot be reached by gas pipes.

Small containers of liquid gas can be used anywhere.

Other types of gas can be turned into a liquid. Propane, a gas taken from oil or natural gas, is stored as a liquid in small containers. Liquid propane can be used to heat up camping stoves or even to power this huge hot-air balloon.

Why can it be dangerous to transport natural gas?

Using natural gas

Natural gas gives off lots of energy when it burns. Energy makes things work. We use energy to heat our homes with gas central heating. We also use gas to heat water for washing and for warm baths and showers.

◀ At home your gas comes from a pipe in the street or from a tank.

In towns and cities, most houses are supplied with gas from 'the mains' – a network of gas pipes under the road. In smaller areas, the network of pipes may not be available. Instead, these houses may have a tank of liquid gas in the garden. This is refilled from a lorry when the gas is running low.

◀ The energy from gas can be used to cook.

When natural gas burns, the energy can be used to heat soup on a gas ring or roast a chicken in a gas oven. A gas metre measures how much gas you use in your home. You then pay for the gas that you use.

The heat energy from natural gas can be turned into electricity.

You use electricity from the moment you wake up and turn on the light to when you go to bed and listen to a CD. **Fossil fuels** are often used to make electricity. In a power station (right), natural gas heats water to make steam which turns a **turbine**. This turns coils of copper wire that pass through magnets, creating electricity.

 How is natural gas used in your home?

Unusual uses of natural gas

We use natural gas to heat our homes and to make electricity in a power station. But there are also other hidden ways that we use this fuel. Helium balloons are filled with a type of natural gas, for example.

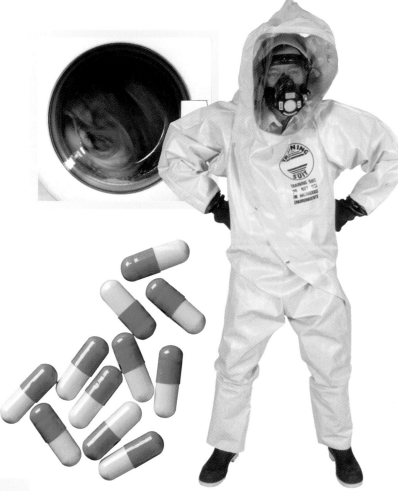

◀ **We use the chemicals in gas to make things.**

Before natural gas reaches our homes, some of its chemicals and gases are taken out. These are made into other chemicals that can be used to make washing powders, plastics and even medicines. Natural gas is used in many areas of our lives!

◀ Natural gas helps crops grow.

Farmers use fertilisers to add goodness to the soil to make crops grow healthy and strong. Some fertilisers are made with sulphur and nitrogen, which come from natural gas.

Gas can also be used to power air conditioners.

Air conditioners are usually powered by electricity. However, gas air conditioners are now being used. These models use less energy and produce fewer gases that harm the atmosphere. Gas is also used to power fridges in areas where there is no electricity – this is useful for hospitals in some developing regions.

 How many gases can you name?

Natural gas and pollution

Each day, we use fossil fuels to drive our cars and to make electricity for computers and lights. These fuels release chemicals into the air that can harm our world. This is called pollution.

◀ **Burning huge amounts of gas creates pollution.**

In some parts of the world, oil is drilled to be used as fuel. Any natural gas released from the same area is left to burn in blazing fires, because it is too costly to store and transport. This is called **flaring**. These raging fires release many chemicals that harm the local area. It would be better if this gas could be stored and then used for fuel.

► Pollution is making the Earth hotter.

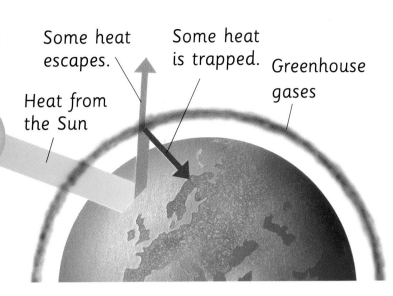

Some heat escapes.

Some heat is trapped.

Greenhouse gases

Heat from the Sun

Fossil fuels release a gas called **carbon dioxide** when they burn. This gas acts like a blanket, trapping the Sun's heat around the Earth. This extra heat can cause extreme weather conditions, from terrible storms to hotter weather.

Natural gas is one of the cleanest fossil fuels.

When gas is burned it does not produce as many harmful chemicals as other fossil fuels. A gas-powered car or bus creates less pollution than a petrol-powered vehicle. Many people are now using this cleaner fuel.

 Why do people prefer to use gas instead of coal and oil?

Natural gas and safety

Natural gas burns easily, which makes it useful for providing energy. But this also means that gas can be dangerous. Great care must always be taken when taking gas from the ground, moving gas from place to place or using gas in the home.

◀ When gas is transported it has to be carried safely.

A single spark can cause gas to burn. Ships that carry gas (called tankers) now have to be specially made so that they are less likely to leak if they crash. Empty spaces in the tankers are also filled with a gas which will not catch fire.

◀ Gas leaks are very dangerous.

If gas escapes and comes near to a flame or other chemicals, it can explode. In 2004, 23 people were killed when a gas pipeline exploded in Belgium. In 1988, an explosion on a drilling rig in the North Sea, called Piper Alpha, also caused many deaths.

Gas smells like rotten eggs!

When natural gas comes out of the ground, it does not smell. A chemical is added to make it smell unpleasant, but very noticeable! If you smell natural gas you should tell a grown-up straightaway. People from a gas company can find the leak and make it safe.

 Why is natural gas made smelly?

Will natural gas last forever?

We are using up natural gas supplies very quickly. One day, we will run out of this form of energy. We need to find alternative sources of energy before it is too late.

◀ Some countries do not use fossil fuels very often. Other countries use them all the time.

In the **industrialised** world, we depend on fossil fuels for energy. Burning fossil fuels causes pollution. Pollution has an effect around the world, causing problems such as extreme weather patterns. Even countries that have limited energy sources are affected by pollution.

▶ We can't reuse or replace natural gas.

Natural gas is a **non-renewable** fuel. When supplies have been used up, it will take millions of years to make any more. However, some fuels, like wood, are renewable. We plant new trees so that there are always supplies of wood.

We are using up fossil fuels very quickly.

In 2003, scientists estimated that, at the rate we are using energy, there is enough natural gas left for about 66 years. Other scientists argue that new discoveries of natural gas deposits mean that we won't run out of this fuel quite so soon.

 What might happen when natural gas runs out?

Gas alternatives

Natural gas from the ground is not the only gas that we can use for fuel. There are other types of gas that we can use. We can also make natural gas from other materials, such as coal, or rotting plants or vegetables.

◀ **Cow manure can be used to make gas.**

Methane is a gas found in cow manure, human waste and rotting rubbish. We burn the methane gas from cow manure to make fuel. The manure is put in special tanks to collect the methane which we call '**biogas**'. This gas can be used for heating and lighting, or to generate electricity.

◀ **Waste dumps create gas that can be used as a fuel.**

Scientists are looking at other ways to make natural gas from rotting waste material. Methane is taken from landfill sites where it collects as the rubbish breaks down. The vegetable remains that we throw away can also be turned into gas.

Gas can be taken from coal.

Heating and refining coal in a power station (right) can remove polluting gases like sulphur. The remaining gases can then be heated to turn gas turbines that produce electricity. There is currently more coal than gas in the world. If gas runs out, some countries may use coal to make energy.

 Why is it a good idea to take methane from waste dumps?

Saving gas today

We can't replace natural gas but we can use less energy. Insulation and double glazing in your home help to save heat. Even just turning off lights when not needed can make a big difference! Saving energy also reduces pollution.

◀ **Recycling can save energy.**

Materials such as glass and plastic can be recycled. This means that they can be broken down and used to make something new. Recycling uses less energy than making objects from new glass or plastic. If you recycle two glass bottles it saves enough energy to boil five cups of tea!

► Turning down the heating can save gas.

In most homes, more than half of the energy used is for heating. You can save gas by turning the heating down slightly and wearing a jumper instead. Boilers that only heat your water when you need it are also a good way to save energy.

Factories could also save a lot more energy.

Gas is widely used in industry. Factories need to reduce the amount of energy that they are using. Some factories are buying more efficient equipment that needs less energy to work. Factories are also looking at ways to reduce the amount of polluting gases that they release into the air.

 How could you and your family save energy?

The future of natural gas

Although new areas of gas are being found from time to time, we are using up our known supplies of natural gas very quickly. Scientists are now looking for alternative sources of natural gas. They are also trying to find fuels that do not cause as much pollution.

◀ There are natural gas supplies deep in the Arctic ice.

This natural gas has been found in wet snow and ice. The Arctic could be a useful new source of natural gas. However, extracting the gas might seriously damage the natural balance of the area.

◀ New ways of drilling for gas do not damage the natural environment.

Powerful laser beams of light can be used to drill into the ground. This causes less pollution than a huge mechanical drill. Instead of building offshore rigs, robotic machines (left) are sent to the bottom of the ocean.

Seaweed can be turned into gas.

Ocean plants, such as sea kelp, can be turned into methane gas that we can burn for energy. Sea kelp farms have been set up in the ocean to grow the seaweed. In the future, huge kelp farms could produce renewable gas energy.

How would a gas pipeline affect life in the Arctic?

See how much you know!

Where is natural gas made?

What does natural gas look like?

What do we use natural gas for?

How do people get natural gas out of the ground?

How does natural gas reach our homes?

How can natural gas cause pollution?

What can we use for energy instead of natural gas?

Why is it important to save energy?

Where are other types of gas found?

Key words

Drill

Coal   **Light**

Energy   **Oil**

Fuel   **Pipe**

Gas   **Pollution**

Heat

Glossary

Biogas – A gas made from animal and human waste, and rotting plant matter. Biogas is mostly methane.

Carbon dioxide – A gas released when fossil fuels burn.

Flaring – The burning of natural gas taken from the ground.

Fossil fuel – A fuel such as coal, natural gas and oil made from the rotted remains of dead animals and plants.

Geologist – A person who studies the Earth and its formation.

Industrialised nation – A country that has developed its industry and economy to create a high standard of living.

Methane – A gas made of carbon and hydrogen that can be used as a fuel.

Non-renewable – Something that cannot be made again.

Reservoirs – Large underground areas of natural gas.

Turbine – A machine used to make electricity.

Index